This book belongs to:

_____

Outer Space

TOPTHAT! Kids™

CONTENTS

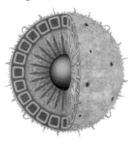

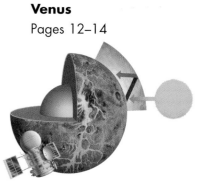

# CONTENTS

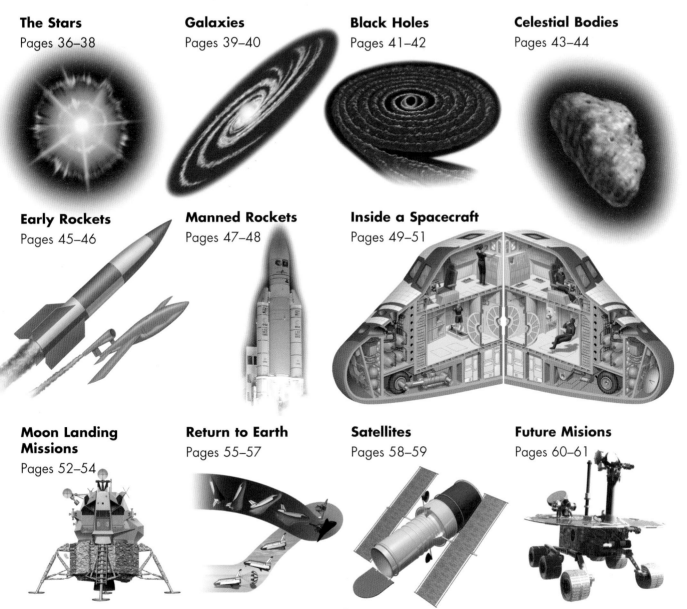

THE UNIVERSE

The universe is a vast, perhaps infinite, mass of galaxies. Older than we can possibly imagine, it is measured in 'light years' due to its phenomenal size. Whether it was formed by a 'big bang', or it developed in another way, scientists have created their own theories about the secrets it may hold ...

How big is the universe ?

The universe contains all space and time, all forms of matter, energy and momentum. It is so big that astronomers have to use the distance that light travels in one year to measure it. They estimate that the universe is at least 93 billion light years across! To give you an idea of how big that is, a beam of light can travel from Earth to the Moon in just one second. That's 300,000 km per second! Just imagine how far light can travel in 93 billion years!

8 minutes

Earth is eight light minutes from the Sun

An artist's impression of the 'big bang'

What was the 'big bang' ?

Most astronomers think that around fourteen billion years ago, the universe arrived with a bang! At that time, the entire universe was inside a bubble that was thousands of times smaller than a letter on this page. It was hotter and denser than anything we can imagine. Then it suddenly exploded, and the universe was born. Nothing existed before this 'big bang' – no stars, no space and no time.

How far can astronomers see into the universe

With powerful telescopes, like the Hubble pictured below, astronomers can see galaxies that are over ten billion light years away, which means the light from them has taken ten billion years to get to Earth. More than

A space observatory

ten billion light years is almost the same distance as a hundred billion trillion kilometres!

Astronomers cannot say for certain what lies beyond this distance. It is impossible to say that the universe definitely stops or ends somewhere – which means that it could even go on forever!

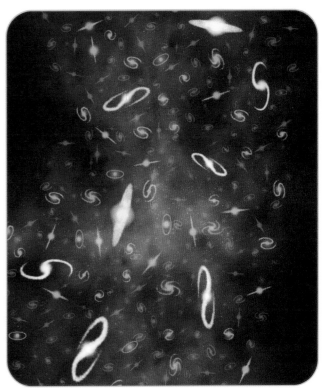

The universe is full of galaxies

The **Hubble Space Telescope**

FACT FILE

There are no certain facts to be had about the past and future of the universe.

Different scientists have different explanations for the way things are and do not always agree.

Most believe in a 'big bang' theory, others have put forward different theories – what do you think?

THE SUN

The Sun makes life on our planet possible by giving us great amounts of light and heat. It is situated at the centre of our solar system and all the planets and other objects orbit around it. Without the Sun, no living thing would be able to survive and our planet would be completely frozen.

Where did the Sun come from ?

Five billion years ago, the part of space where our solar system now exists was full of clouds of hydrogen gas and dust. Over billions of years, this gas and dust slowly moved together, due to gravity. As more and more gas and dust came together, nuclear reactions began to take place and the gas started to shine!

The Sun orbits the centre of the Milky Way galaxy. It is made up of around 75 per cent hydrogen and 25 per cent helium. It measures more than a million kilometres across – so big that you could fit more than 1 million Earths inside it!

FACT FILE

The corona is the outermost layer of the Sun. It stretches millions of kilometres into space.

Believe it or not, the Sun is just a star, just like those we see twinkling at night.

How hot is the Sun ?

At its centre the Sun is an extreme 15 million°C – so hot that planets millions of kilometres away receive its heat! The Sun's temperature slowly decreases towards its surface where it is about 6,000°C. This is cool for the Sun, but is actually about 16 times hotter than boiling water! At the outermost layer, something strange happens because the temperature rises again to well over 1 million°C!

Sir Arthur Eddington

During the first half of the twentieth century, Sir Arthur Eddington explained that heat and light is generated by the Sun when particles called protons crash into the Sun's core.

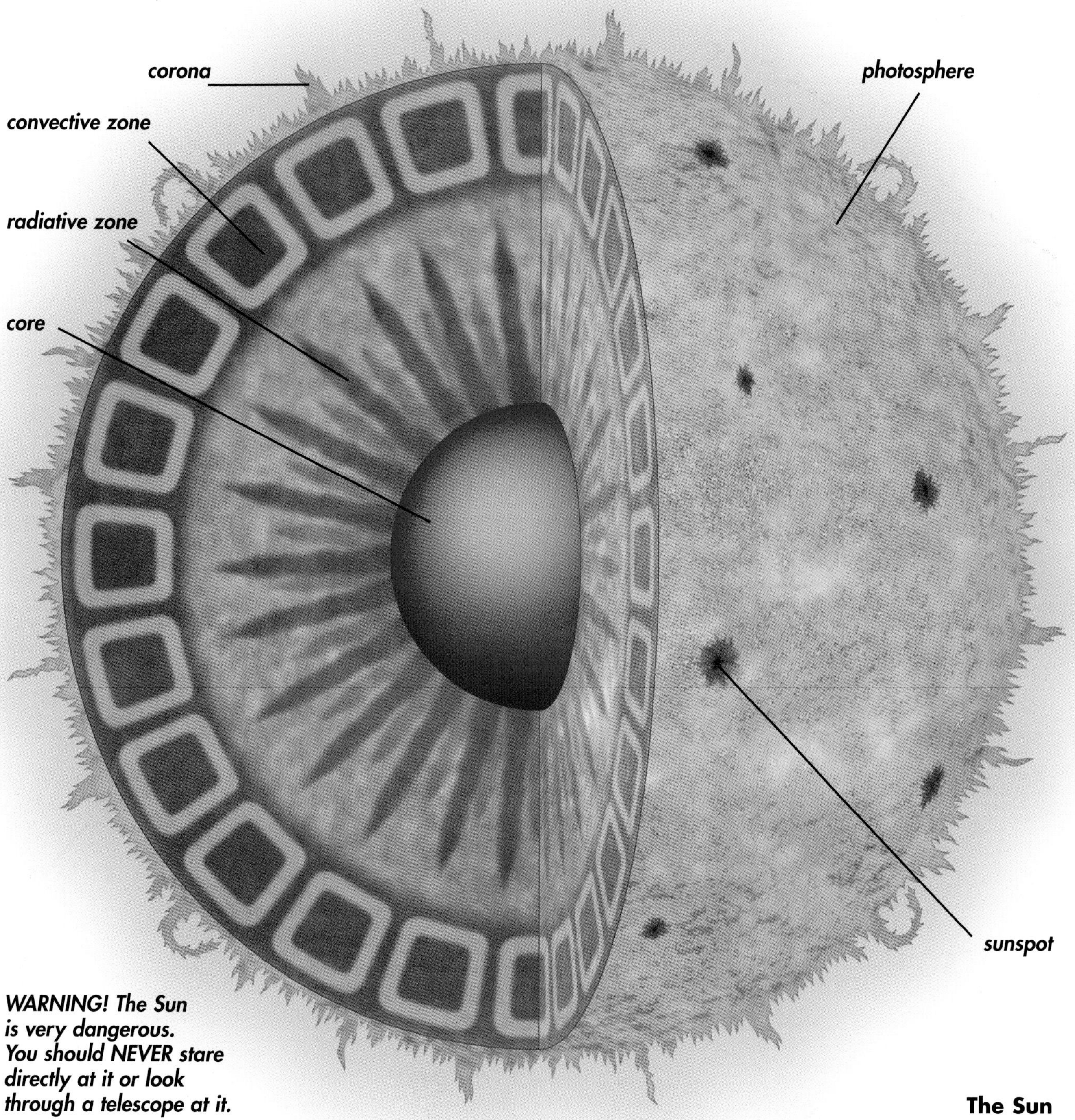

corona

convective zone

radiative zone

core

photosphere

sunspot

WARNING! The Sun
is very dangerous.
You should NEVER stare
directly at it or look
through a telescope at it.

The Sun

7

Why do we have seasons ❓

We have seasons – spring, summer, autumn and winter – because Earth is tilted. This tilting causes different parts of the globe to be positioned towards the Sun at different times of the year. If the northern hemisphere is tilted towards the Sun, it will be summer there. At the same time, the southern hemisphere will be tilted away from the Sun and it will be winter. Autumn and spring occur when Earth is tilted neither towards or away from the Sun. This means different sides of the world experience opposite seasons at the same time.

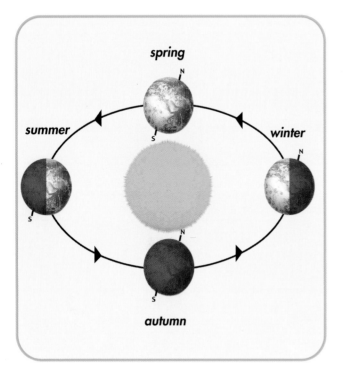

The seasons in the northern hemisphere. Note the tilt in relation to the Sun

What is a sunspot ❓

Sunspots are strange and powerful phenomenon. Sometimes as big as Earth, they move across the surface of the Sun, shifting in size and shape as they go. Cooler and darker than surrounding areas of the Sun's surface, sunspots occur when a magnetic field is formed below the Sun's surface.

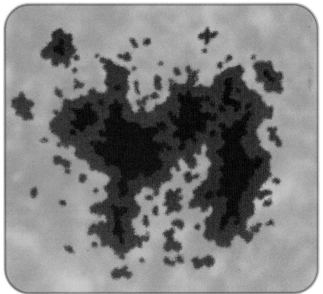

Enlarged area of the Sun showing a sunspot

FACT FILE

If you add up all of the mass in the solar system, including the planets, the moons, the asteroids, the comets, the dust, you and everything else, it turns out that 99.85 per cent of everything is the Sun.

MERCURY

Mercury is the nearest planet to the Sun and the second smallest in the solar system. It has the widest range in temperature of any planet, from a chilly -180°C at night to a scorching 430°C during the day. If you were standing on Mercury, the Sun would look three times larger than it does from Earth!

How big is Mercury ?

With a diameter of 4,880 km, Mercury is around 30 per cent the size of Earth. It orbits the Sun every 88 days. Mercury is similar in appearance to the Moon. It has lots of impact craters, no moons and no large atmosphere.

The respective sizes of Earth (left) and Mercury (right)

How hot is Mercury ?

430°C

-180°C

Mercury experiences extreme temperature variations

Although Mercury is the planet closest to the Sun, Venus, the third planet, is slightly hotter because it has an atmosphere that traps more heat. Mercury has no atmosphere, but has very extreme temperatures, ranging from 430°C in sunlight to -180°C at night.

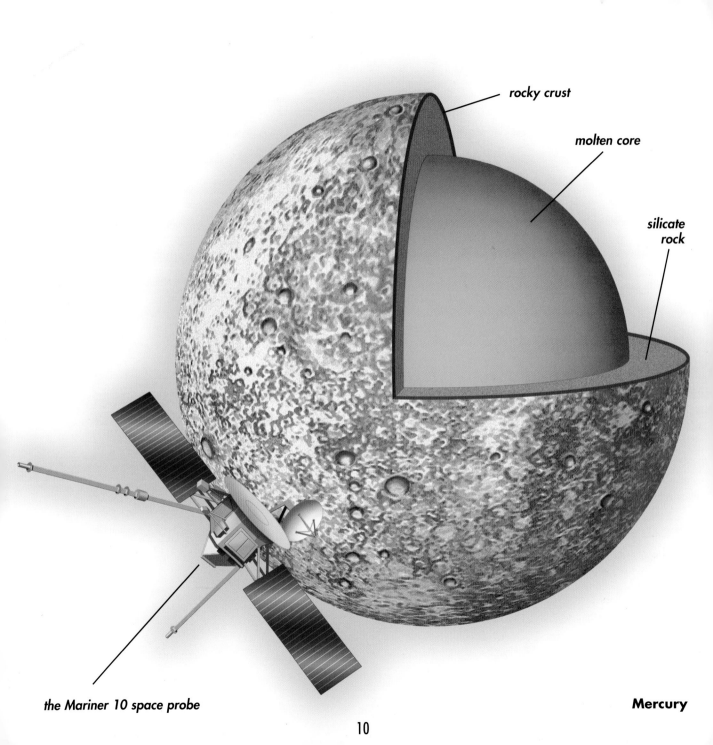

rocky crust

molten core

silicate rock

the Mariner 10 space probe

Mercury

How long is Mercury's year

One year on Mercury is equal to 88 Earth days. However, a day on Mercury is almost twice as long as a year! If you were to stand on the surface of Mercury, the time from one sunrise to another, would be equal to 176 Earth days. These long days and nights cause temperatures to rise very high and fall very low.

How many times has it been visited

Two space probes have visited Mercury. The *Mariner 10* was launched in 1973 and flew past Mercury in 1974, collecting the first ever images and data of the planet. The *MESSENGER* was launched in 2004, with the first of three flybys occurring in 2008.

Why is it called Mercury ?

The planet was named by the Romans in honour of Mercury, messenger of the gods. Mercury wore a hat and sandals with wings on them, allowing him to travel around the world very quickly. The planet was named after him because it moves around the Sun faster than any other planet in the solar system.

FACT FILE

Because of Mercury's weird orbit and rotation, the morning Sun appears to rise briefly, set and rise again. The same thing happens in reverse at sunset!

Mercury is roughly the same age as the Sun – 4.5 billion years old!

Can Mercury be spotted in the night sky ?

Mercury is not usually visible from Earth as the two planets orbit the Sun at different angles. On rare occasions, however, Mercury passes directly between the Sun and the Earth. When this happens Mercury can be seen as a small spot on the Sun's surface.

This event takes place about thirteen or fourteen times each century. The last time was on 8 November 2006 and the next time will be on 9 May 2016.

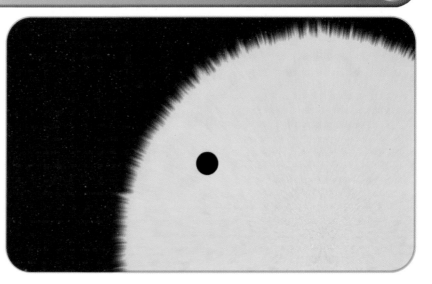

Mercury crossing over the sun

VENUS

Venus is the second-closest planet to the sun. It is the brightest object in the night sky, after the Sun and the Moon. It is the only planet where the Sun rises in the west and sets in the east. Under Venus's bright white clouds, exists a cratered surface on which nothing could survive.

Why is Venus so bright

It is partly due to its covering of white clouds, which reflect 60 per cent of the Sun's rays back into space. Astronomers measure the brightness of a star on a 'magnitude' scale. Venus can reach a magnitude of -4, making it one of the brightest objects in the night sky.

An artist's impression of Venus as seen from Earth

Has a spacecraft ever landed on Venus

Spacecraft have landed and orbited Venus since the 1970s. One of the first to land was a Russian spacecraft, *Venera 9*, in 1975. It collected data and pictures of Venus and sent them back to Earth.

How close is Venus to Earth

Venus is the nearest planet to Earth. It comes within 38 million kilometres of our planet. Venus is sometimes described as Earth's twin, as they are so similar in size, mass and their distance from the Sun. However, that is where the similarities end, as the conditions on Venus are very different to those on Earth. Venus can be seen from Earth several times a year, sometimes in the evening after sunset or in the morning before sunrise.

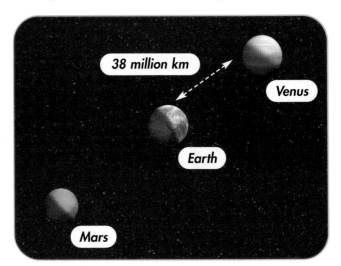
Venus' orbit takes it relatively close to Earth

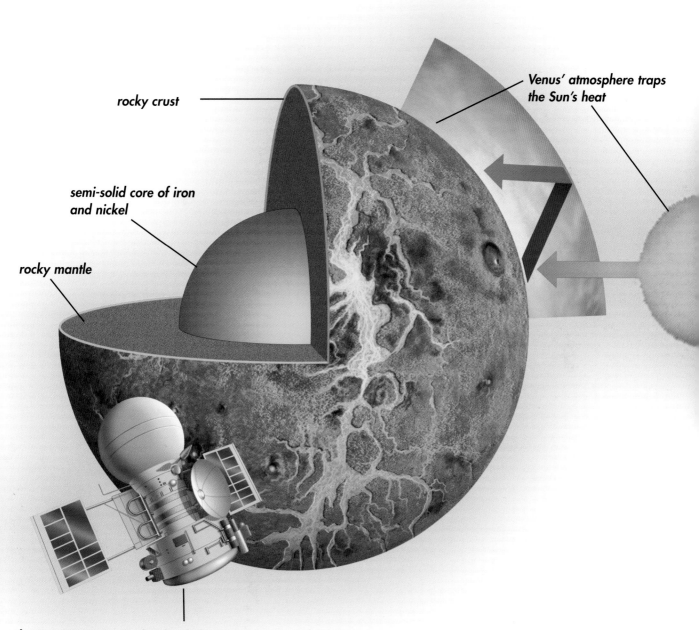

rocky crust

Venus' atmosphere traps the Sun's heat

semi-solid core of iron and nickel

rocky mantle

the Russian Venera 9 that landed on Venus

Venus

How hot is Venus

Venus is the hottest planet in the solar system. The thick clouds and the atmosphere, which is mostly made up of carbon dioxide, makes Venus just like a greenhouse, trapping in the Sun's heat. The temperature on its surface can reach 470°C. Venus retains a constant temperature between day and night and between the equator and the poles. Astronomers think it is unlikely that life exists on Venus, due to the intense heat.

FACT FILE

Venus probably once had large amounts of water like Earth but it all boiled away. If Earth was any closer to the Sun, it may have suffered the same terrible fate as Venus.

There are no small craters on Venus. Small meteoroids burn up in Venus' dense atmosphere before reaching the surface.

How long is a year on Venus

A year on Venus is shorter than a year on Earth. It takes Venus 225 Earth days to orbit the sun. If you were on the surface of Venus, the time from one sunrise to the next would be 117 Earth days. You would see the sun appear in the west and set in the east – the opposite to Earth. This happens because Venus orbits the Sun in the opposite direction to Earth.

What is it like on the surface of Venus

The dusty surface of Venus is littered with impact craters and volcanoes. Thick clouds of sulphur dioxide, capable of producing lightning much like the clouds on Earth, permanently fill the skies. Its gentle rolling plains are covered in gloomy darkness, caused by the dense clouds that reflect about 60 per cent of the sunlight that falls on them back into space.

The surface of Venus could be volcanic

EARTH AND MOON

Our planet is unique, with an atmosphere of gases existing in perfect harmony to sustain life. The planet is the correct distance from the Sun for water to exist as a liquid, a vital ingredient for life. The Earth's tides are affected by the Moon, formed billions of years ago.

Why does Earth have a moon ?

Astronomers believe that billions of years ago, Earth collided with another planet, known as the 'Giant Impact' or 'Big Whack'. As a result of the impact, lots of rock was blasted into space. Over time this rock reformed as the Moon. The Moon is much smaller than Earth and has a diameter of 3,470 km, compared to Earth's 12,700 km. The oldest rocks on the surface of the Moon are estimated to be 4.6 billion years old.

The Moon is the result of Earth's collision with another planet

How old is Earth ?

Scientists estimate that Earth formed 4.54 billion years ago out of the mass of dust and gas left over from the formation of the Sun. Just as the Sun and planets were born, so they will eventually die. As the Sun ages, it will enlarge and consume some of the planets, possibly even Earth, before shrinking to become a dense white dwarf. When the Sun dies, the orbits of the remaining planets will be thrown into chaos, causing some to collide and others to be thrown into space.

FACT FILE

The combination of gases which make up our atmosphere: 78 per cent nitrogen, 21 per cent oxygen and 1 per cent other gases.

Humans haven't existed for very long at all – primitive life began here about 200,000 years ago, but the oldest rocks discovered so far in Earth's crust are about 3,900 million years old!

Earth and the Moon

What makes Earth suitable for life

Earth is only a tiny part of the universe, but is home to all known life forms. It provides the ideal amount of water and the environment for life as we understand it to flourish. The distance of Earth from the Sun, as well as its path around the Sun, tilt, rotation, atmosphere and protective magnetic field all contribute to the conditions necessary to sustain life on this planet.

Earth is home to all known life in the universe

Why does the Moon appear to change shape

As the Moon orbits Earth, it appears to change shape in the sky. The bright part of the moon that we can see changing shape from Earth is called its 'phase'. The phase of the Moon depends on its position in relation to the Sun and Earth.

The phase varies from a full moon (when Earth is between the Sun and the Moon) to a new moon (when the Moon is between the Sun and Earth). The Moon orbits Earth every four weeks, meaning there is just over twenty-nine days between full moons.

Some of the phases of the Moon

What is an eclipse

When the Moon comes between the Sun and Earth, the light reaching Earth will be affected. This is called a 'solar' eclipse. If they cross slightly, then we call it a 'partial' solar eclipse. If our light is totally blocked, it is known as a 'total' eclipse and everything goes dark.

When Earth passes between the Sun and the Moon, it is known as a 'lunar eclipse'. The Moon will turn a dark red, as sunlight passing through our atmosphere is 'bent' or scattered onto the Moon.

A total eclipse

MARS

Mars is the fourth planet from the sun in the solar system. Its mysterious red surface, littered with impact craters, mountains, canyons and volcanoes has captivated astronomers for hundreds of years. It is home to the highest mountain and the deepest canyon in the solar system.

What is it like on Mars

The surface of Mars is dusty, rocky and red. There are no oceans, streams or rivers. Its northern plains are barren and flat, while the mountainous south is pitted and cratered by ancient impacts. Temperatures can drop to -140°C during winter and rise to 20°C in summer. The largest dust storms in the solar system rage across the entire planet. Mars is home to the highest mountain known to man, Olympus Mons, which stands three times higher than Mount Everest.

Could life exist on Mars

For hundreds of years people speculated that life could exist on Mars. However, when the first space probe flew past Mars in 1965, it found that life was unlikely to exist there. Its reports of a cratered, seemingly dead world shook the scientific community. Since then, however, scientists have found strong evidence to suggest that water once flowed on the surface of Mars, and simple life may once have existed there.

Why is Mars red ?

Mars is red because its soil is rich in iron oxide. Some astronomers think that at least some of the planet's iron came from meteorites – a theory supported by the fact that Mars' surface is covered with impact craters. The 'blood-like' colour is one reason why it is named after the Roman 'God of War'.

An artist's impression of the surface of Mars

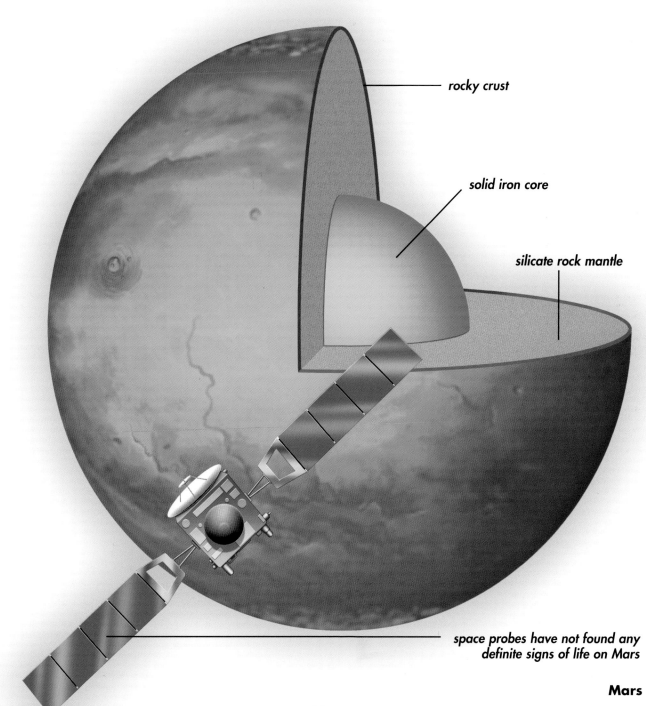

rocky crust

solid iron core

silicate rock mantle

space probes have not found any
definite signs of life on Mars

Mars

19

Does Mars have a moon

Mars has two tiny moons, called Phobos and Deimos, which are asteroids that have been 'drawn in' or 'captured' by Mars' gravity. An asteroid is a rocky object in space. There are hundreds of thousands of asteroids within the main asteroid belt between Mars and Jupiter.

Thousands of asteroids are discovered each year

Asteroids vary greatly in size, from 1 km up to 200 km across

FACT FILE

Mars has a massive volcanic mountain called Olympus Mons, which is 27 km high and 600 km across!

Its volume is about 100 times larger than Earth's largest volcano.

What are 'Martian canals'

In the late 1800s and early 1900s, various astronomers reported seeing strange canals on the Martian surface, through telescopes. The most famous of these astronomers was Percival Lowell, who built a huge telescope in Arizona to study Mars.

In an age when people were inspired by voyage and discovery, many people were convinced by his reports. However, by the end of the twentieth century, it was obvious that Mars might only be able to support primitive life. The canals were not built by intelligent beings – they had been a trick of the eye and the blurring effect of Earth's atmosphere.

Mars' surface appears red to us because it contains iron oxide dust. The 'blood-like' colour is one reason why it is named after the Roman 'god of war'

JUPITER

Jupiter has many interesting features, including its famous 'Great Red Spot'. With an incredible gravitational pull and turbulent clouds, it is the largest planet in the solar system, and the fifth furthest away from the Sun.

How big is Jupiter

Jupiter is gigantic. It has an equatorial diameter of 142,000 km, making it eleven times wider than Earth. In fact, you could fit about 1,300 Earths inside Jupiter with room to spare! Jupiter is two and a half times as massive as all of the other planets in our solar system combined.

What is the Great Red Spot?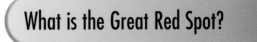

The Great Red Spot is a swirling mass of gas, found in Jupiter's southern hemisphere. Resembling a gigantic rotating storm, the widest part of the spot is about three times larger than the whole of Earth. The storm is so large that it has perhaps raged for as long as 343 years or more.

How far is Jupiter from the sun

Jupiter is about 780 million kilometres from the Sun – around five times further away than Earth. Light, travelling at 300,000 km per second, takes over 40 minutes to get to Jupiter. It is so far away, it takes nearly twelve Earth years to orbit the Sun. Jupiter's rotation is the fastest of any planet in the solar system. Because it is not a solid body, the rotation of its poles is about five minutes longer than that of its equator.

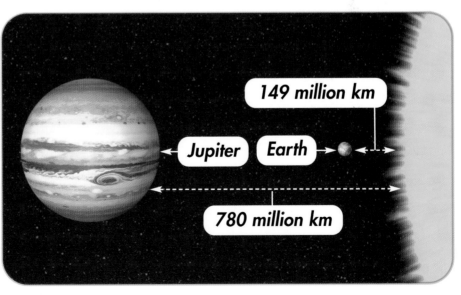

149 million km

Jupiter Earth

780 million km

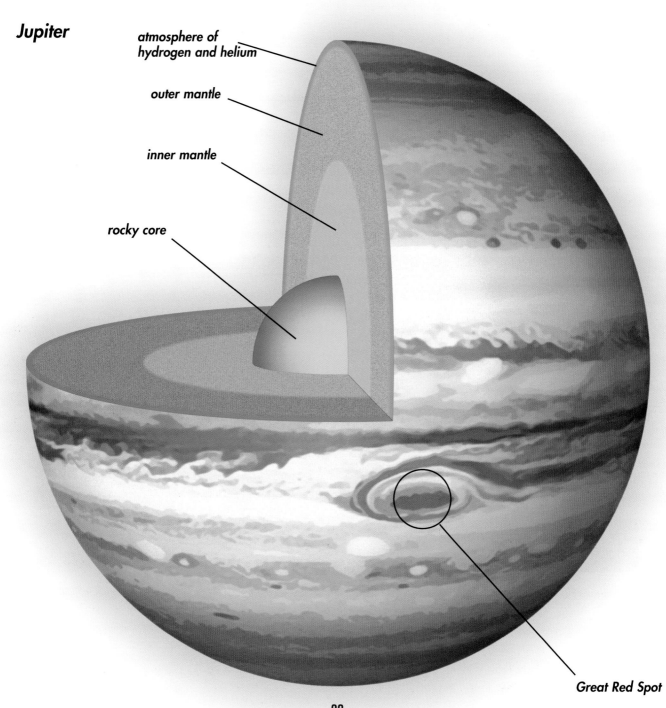

Jupiter

atmosphere of
hydrogen and helium

outer mantle

inner mantle

rocky core

Great Red Spot

22

How many moons orbit jupiter

Jupiter has over sixty natural satellites, or moons. The planet has such a high number of moons because over billions of years it has captured more and more objects in its huge gravitational field. The four largest moons, known as the 'Galilean moons', are called Io, Europa, Ganymede and Callisto. They are each a similar size to Earth's Moon. These four moons can be spotted as faint stars through a small pair of binoculars. In July 1994, Jupiter's gravity pulled the comet Shoemaker-Levy 9 into the giant planet. The impact caused huge explosions, and the scars from the collision could be seen on Jupiter for many months after the impact.

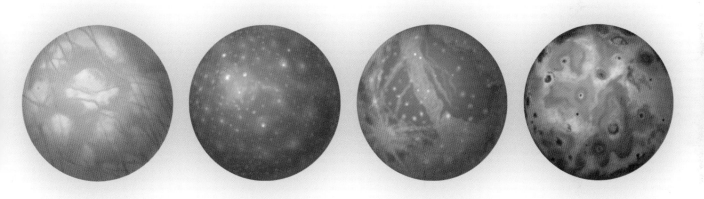

Europa
Diameter 3,138 km

Callisto
Diameter 4,800 km

Ganymede
Diameter 5,262 km

Io
Diameter 3,630 km

Is it possible to land on Jupiter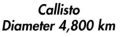

Sending a manned spacecraft to Jupiter would be extremely challenging. Along with Saturn, Uranus and Neptune, Jupiter is known as a gas giant because it is extremely big and has no solid surface! If man were to travel to Jupiter, it is likely that it would be a mission to one of Jupiter's moons, rather than the planet itself.

Does life exist on Jupiter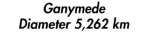

The high pressure of Jupiter's gases and the small amount of water in its atmosphere makes it impossible for life as we know it on Earth to exist there. Some scientists predict, however, that floating creatures, called atmospheric beasts, may exist in Jupiter's upper atmosphere.

SATURN

With its massive spinning ring system, consisting mostly of ice particles with a smaller amount of rocky debris and dust, Saturn is one of the most beautiful objects in the night sky. A little smaller than Jupiter, it is the sixth planet from the Sun and the second largest in our solar system.

Why has Saturn got rings ?

Astronomers once thought that Saturn's rings were the result of an icy moon that came too close to Saturn and broke up. However, recent research has shown that the particles in the rings differ in age, making some astronomers believe that the rings weren't created during a one-off event but that material is constantly coming together to form 'moonlets', which then break up again, creating a recycling process.

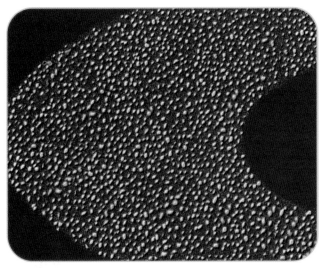

Did an icy moon break up to form Saturn's rings?

How big are Saturn's rings ?

Saturn's magnificent rings span a distance of over 300,000 km, with an average thickness of around 1 km. They are made up of millions of icy particles, with some combinations of dust and other chemicals. The particles range in size from a grain of sugar to the size of a house! The rings can be viewed from Earth using a modern telescope.

Saturn's rings are made from icy chunks

Do other planets have rings ?

Saturn is not the only planet to have a system of rings. The other gas giants, Jupiter, Uranus and Neptune, have extremely feeble ring systems, which are too faint to be seen from Earth.

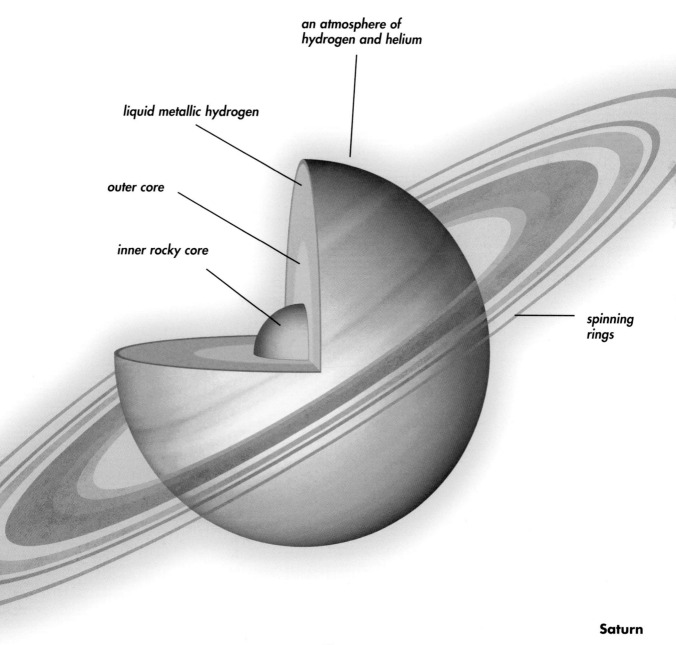

an atmosphere of
hydrogen and helium

liquid metallic hydrogen

outer core

inner rocky core

spinning
rings

Saturn

Can Saturn be spotted in the sky

Saturn can be viewed with the naked eye, but without the aid of a telescope, it can be tricky to spot against the background of stars. The give-away sign of any planet is that it doesn't 'twinkle' like stars do. If you view Saturn through a telescope, you may be able to see its beautiful rings and make out its distinctive bulging equator.

How large is Saturn

Saturn is the second largest planet in the solar system. It has an equatorial diameter of over 120,000 km. It is thought that its interior is similar to that of Jupiter, having a small rocky core surrounded mostly by hydrogen and helium. Like Jupiter, Saturn is about seventy-five per cent hydrogen and twenty-five per cent helium with traces of water and other elements.

What is Saturn's largest moon

Saturn has dozens of orbiting moons ranging in size, shape and age. The largest of its moons is Titan, which is over 5,000 km in diameter and is the only natural satellite in the solar system to have its own atmosphere. It orbits Saturn around once every sixteen days.

At 5,150 km in diameter, Titan is Saturn's largest moon

How far is Saturn from Earth

The sixth planet from the Sun, Saturn is an average of 1,300 million kilometres from Earth. It takes almost an hour and a half for radio signals, which travel at the speed of light, to reach Saturn if sent from Earth. Saturn is over 1,400 million kilometres from the Sun, almost twice as far as Jupiter, and takes twenty-nine Earth years to orbit the Sun.

Jupiter has a diameter of around 142,000 km

Saturn has a diameter of around 120,000 km

FACT FILE

Saturn spins so fast that forces cause the equator to bulge outwards. It is 10 per cent bigger in the middle than at the poles.

This planet's density is so small that it would float on water!

URANUS

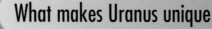

With its blue-green appearance, Uranus must have been an exciting discovery for astronomers. Its distinctive features include a strange rotation system and many moons. It has a narrow ring system, but unlike those of Saturn which are pale in colour, they are dark, as they are made from dark-coloured dust particles.

How was Uranus discovered ?

Uranus was discovered on 13 March 1781 by William Herschel, using a hand-made telescope in the back garden of his house in Bath, England. The house has now been made into a museum. Herschel thought Uranus was a comet when he first saw it. In 1787, he discovered two of Uranus' moons – Titania and Oberon.

Herschel discovered Uranus from his garden!

You couldn't breathe on Uranus as methane makes the atmosphere poisonous

What makes Uranus unique ?

Uranus is the only planet in the solar system to spin on its side! It spins in the same direction as it travels, rolling around its orbit. No one knows for certain what happened to Uranus to make it like this. However, some scientists believe that a massive collision, billions of years ago, may have caused this 98 degree tilt.

98°

Uranus' tilt means it also has a strong magnetic field

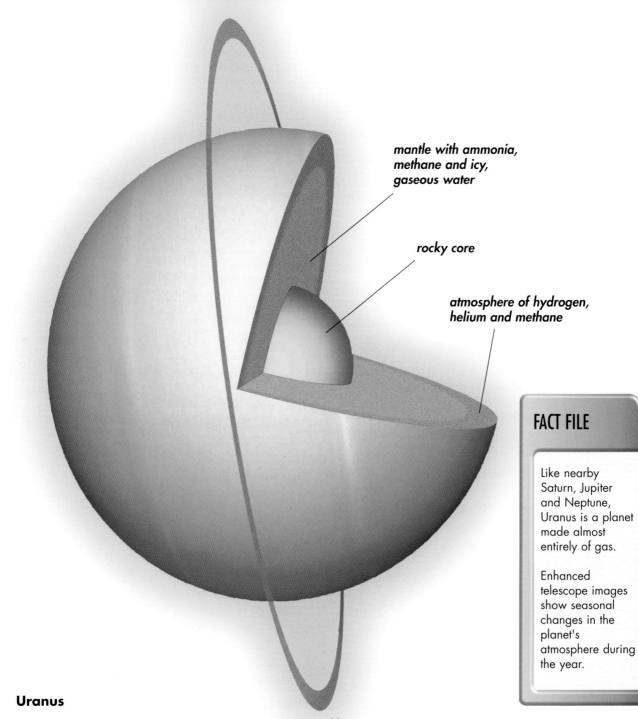

mantle with ammonia, methane and icy, gaseous water

rocky core

atmosphere of hydrogen, helium and methane

FACT FILE

Like nearby Saturn, Jupiter and Neptune, Uranus is a planet made almost entirely of gas.

Enhanced telescope images show seasonal changes in the planet's atmosphere during the year.

Uranus

How many moons orbit Uranus

Uranus has over twenty moons. The four largest ones are called Ariel, Umbriel, Titania and Oberon. Another moon, called Miranda, is less than 500 km in diameter and has a strange grooved and cratered surface, unlike any other moon in the solar system.

With its carved surface, Miranda is unique. It could have been shattered by another moon crashing into it, reforming with a crazy jagged surface

Ariel
Diameter 1,160 km

Umbriel
Diameter 1,170 km

Titania
Diameter 1,580 km

Oberon
Diameter 1,550 km

How can Uranus be spotted

Uranus can be seen as a faint star with the naked eye from a very dark site. A pair of binoculars or a small telescope is needed to see the planet easily. With a large telescope, it appears as a greenish disc. It can be seen all year round in the constellation of Aquarius and is best seen in August.

How far is Uranus from the Sun

Uranus is about 2,900 million kilometres from the Sun. That's twice as far away as Saturn! Light from the Sun takes nearly three hours to reach Uranus. At over 50,000 km in diameter, Uranus is a giant planet, but it is still less than half the diameter of Saturn.

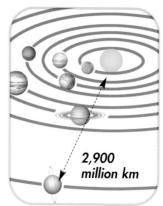

2,900 million km

NEPTUNE

Neptune is the fourth largest planet in the solar system, and the furthest from the sun. It is home to the fastest winds in the solar system, which power across this vivid blue planet at speeds of over 2,000 km/h. It has thirteen moons and a spectacular system of rings.

How was Neptune discovered ?

Neptune has the honour of being the first planet to be predicted on paper before it was actually seen. In 1845–46, French mathematician, Urbain Le Verrier, predicted the existence of the planet using Newton's laws of motion. Then, in September 1846, guided by Le Verrier's calculations, Johann Gottfried Galle and Heinrich d'Arrest of the Berlin Observatory spotted the elusive planet. It was discovered within 1° of where Le Verrier had predicted it would be. The planet was named Neptune after the Roman god of the sea.

Astronomers, Galle and d'Arrest

How long does it take Neptune to orbit the sun ?

It takes Neptune 165 Earth years to orbit the Sun. At a distance of 4,500 million kilometres, it is thirty times further from the Sun than Earth. For this reason, Neptune is extremely cold. The temperature at its equator is about –230°C.

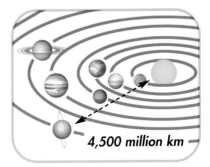

4,500 million km

FACT FILE

Neptune has three main rings, which vary in thickness. This was confirmed by the *Voyager 2* space probe, in 1989.

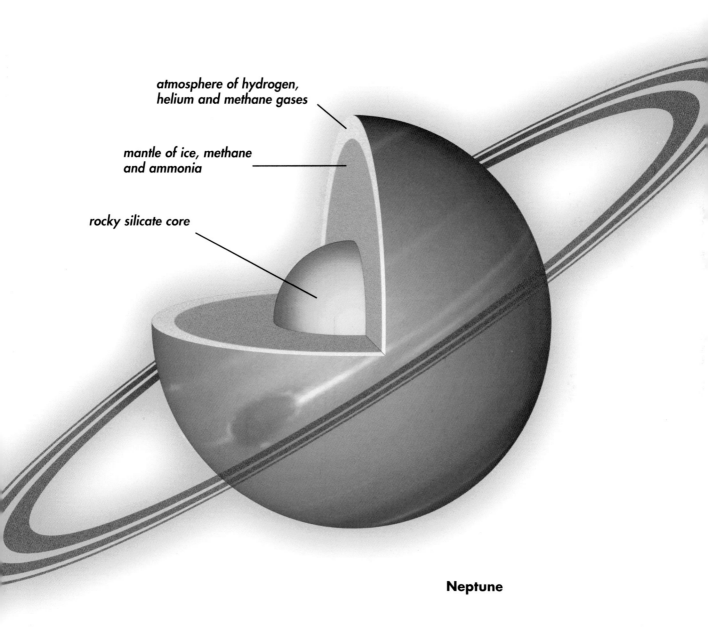

atmosphere of hydrogen, helium and methane gases

mantle of ice, methane and ammonia

rocky silicate core

Neptune

What is Triton ❓

Triton is Saturn's largest moon. It is the only moon in the solar system that has its own dense atmosphere, and the only place other then Earth where clear evidence of stable bodies of surface water have been discovered. Seasonal weather patterns, wind and rain have created on Triton surface features similar to those on Earth, such as shorelines and sand dunes. This has led some scientists to speculate that conditions on Triton resemble those of early Earth, although at a much lower temperature.

The surface temperature of Triton is a brutally cold -235°C

Why is Neptune blue ❓

Neptune's rich blue colour is a mystery. No one knows the true identity of the chemicals that give the clouds their vivid blue tint. However, the blue colour can be partly explained by the methane clouds in the planet's upper atmosphere which absorb red light and reflect blue light.

Neptune's ocean of liquid methane and ice

How is Uranus similar to Neptune ❓

Sometimes called 'planetary twins', these two icy, gaseous planets are both about 50,000 km in diameter and coloured blue-green. Neptune rotates on its axis every sixteen hours, while Uranus rotates on its axis every seventeen hours.

Neptune (left) looks similar to Uranus (right)

PLUTO

Formerly the smallest planet in the solar system, Pluto was reclassified as a 'dwarf planet' after the International Astronomical Union changed the meaning of the term 'planet', in 2006. This remote ball of ice has three known moons. Charon, its closest moon, is about half the size of Pluto itself.

Who discovered Pluto

In 1905, Percival Lowell discovered that the force of gravity from some unknown object was affecting the orbits of Uranus and Neptune. In 1915 he predicted the existence of a new planet, but died in 1916 without finding it. In 1930, Clyde Tombough, using the predictions made by Lowell, found Pluto. Named after the god of the underworld, Pluto also honours Percival Lowell, whose initials are the first two letters of Pluto.

An artist's impression of Pluto's surface

Is Pluto a real planet

In 2006, the International Astronomical Union decided that Pluto should be reclassified as a 'dwarf planet'. The decision was made by astronomers who argued that Pluto shared its orbit with too many asteroids to be considered an independent planet. After the reclassification, Pluto was added to the list of minor planets and given the number '134340'. There are two other known dwarf planets in our solar system. They are called Eris and Ceres.

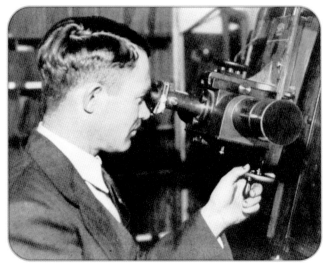

Clyde Tombaugh

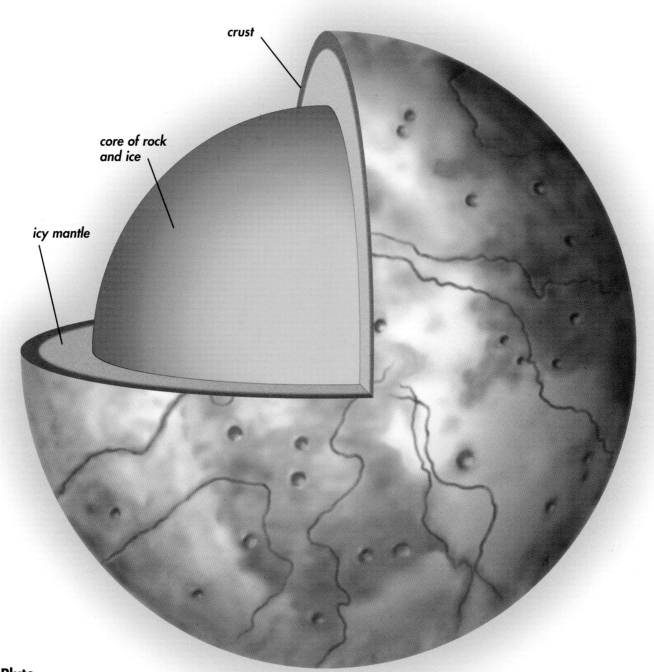

crust

core of rock
and ice

icy mantle

Pluto

How far is Pluto from the Sun

Pluto's orbit is different from those of the planets. While the planets have a circular orbit, Pluto's orbit is highly eccentric, which means that at certain points of its orbit Pluto is closer to the Sun that at other times (Pluto's orbit takes it to within 4.34 billion kilometres of the Sun and as far away as 7.4 billion kilometres).

This means that Pluto is sometimes closer to the Sun than Neptune. The last time this happened was between 1979 and 1999.

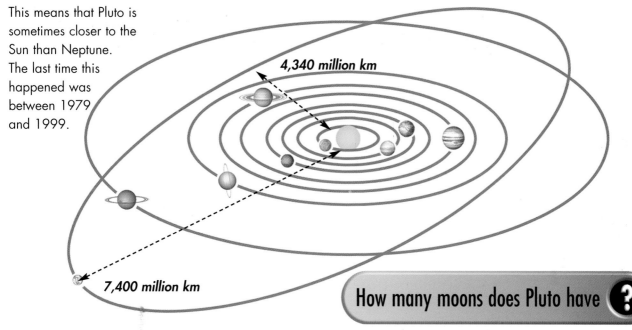

4,340 million km

7,400 million km

How many moons does Pluto have

Pluto has three known Moons. They are called Hydra, Nix and Charon. Discovered in 1978, Charon is Pluto's largest and nearest moon. Pluto and Charon are tidally locked to each other. This means that they continuously face each other as they travel around the Sun. If you were standing on Pluto's near side, Charon would seem to hover in the sky without moving.

Charon is half the size of Pluto, and 19,600 km away.

FACT FILE

Astronomers estimate that the surface temperature of Pluto is about -225°C.

From the surface of Pluto, the Sun is so tiny it looks like a bright star in the sky.

Pluto is tiny – only two-thirds the size of our Moon!

THE STARS

Stars are huge fireballs of gas that can be seen twinkling in the night sky. Our Sun is just one of an estimated 100 billion billion stars in the universe. All the chemicals in your body, from the calcium in your bones to the zinc in your hair, were originally forged in the fiery furnaces inside stars!

How far away are the stars ❓

There are billions of stars in the universe. The closest star, apart from the Sun, is over four light years away. Stars that can only be seen through a telescope are many thousands of light years away. Due to the speed of light and the time it takes for light to travel to Earth, we see stars as they looked several years ago, and faraway stars as they looked thousands of years ago.

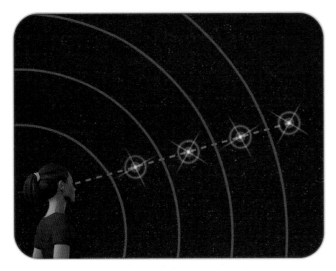

With the naked eye, we can see stars four light years away

Why are stars different colours ❓

If you look carefully at the stars with binoculars or a telescope, you will see stars range in colour, from red to yellow, to blue. The different colours depends on the temperature of the star. Measured in Kelvins, red stars have surface temperatures that range between 2500-3500^K. Yellow stars are about 5500^K and blue stars range from 10,000 to 50,000^K.

The colours of a star depends on its surface temperature

What is a constellation ❓

A constellation is simply a set of stars that have been put into a group that form a recognized pattern, and given a name. There are eighty-eight groups of stars that can be called constellations. Many of them were named after myths and legends.

What are the 'hemispheres' ?

Earth is split around the middle by an imaginary line called the equator. Whether you are situated above or below it determines whether you're in the northern or southern hemisphere.

The stars you can see depends on where you are on Earth. There are some stars which can only be seen from the northern or the southern hemisphere. For example, people in the southern hemisphere can't see the Pole Star (Polaris).

northern hemisphere

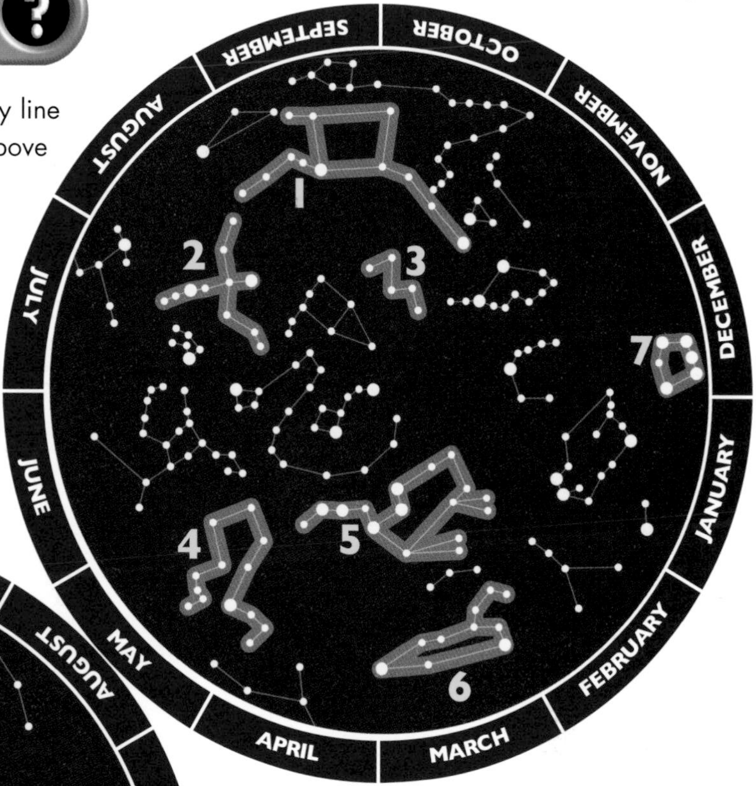

southern hemisphere

Spot some famous star constellations in the northern hemisphere:
1) Pegasus, 2) Cygnus, 3) Cassiopeia,
4) Bootes, 5) Ursa Major, 6) Leo,
7) top of Orion

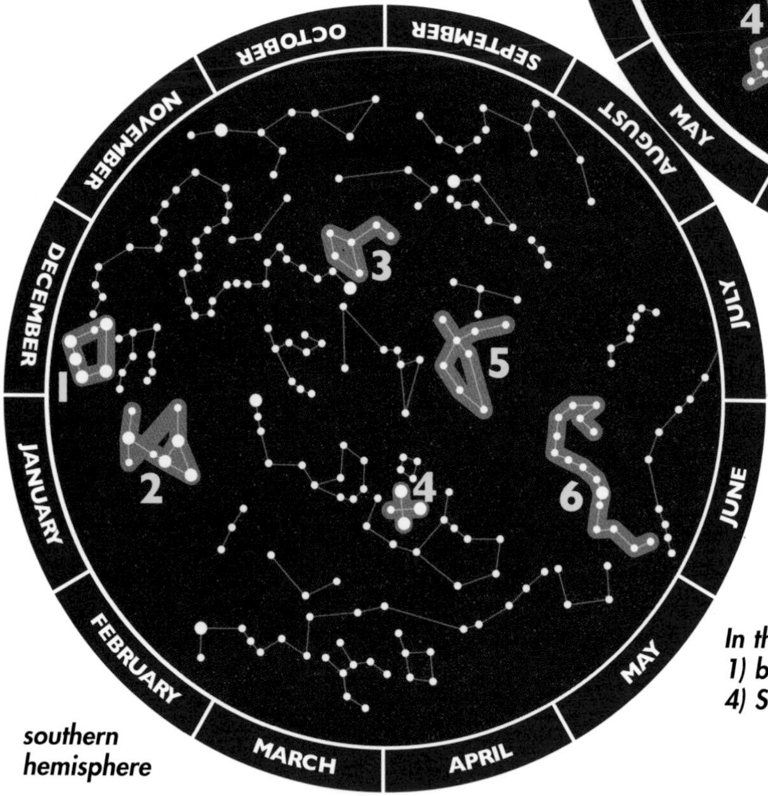

In the southern hemisphere, you can see:
1) bottom of Orion, 2) Canis Major, 3) Phoenix,
4) Southern Cross, 5) Pavo, 6) Scorpios

What is a 'nova'

A nova is a sudden brightening of a star. A nova may occur in a stellar system that has two stars – a white dwarf (sleeping star) and another star. If these two stars are close enough to each other, material from one star can be pulled off its surface and onto the white dwarf. Occasionally, the temperature of this new material on the surface of the white dwarf may become hot enough to reignite the sleeping star. This causes the white dwarf to suddenly become very bright – a nova!

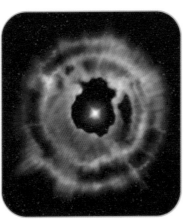

A nova is 100,000 times brighter than a normal star

A star reaches the end of its life

How long do stars last

How long a star lives depends upon its mass. The bigger they are, the quicker they die. This might seem odd, but the more mass a star has, the hotter it gets. The hotter it gets, the quicker it exhausts its fuel supply. Our Sun has billions of years of useful life left before it runs out of hydrogen fuel. It is around half-way through its 'life' as a star.

Our star will burn for billions more years!

FACT FILE

Our Sun will eventually form a white dwarf – a spherical diamond the size of the Earth! From this point on the Sun will gradually fade away, becoming dimmer and dimmer until its light is finally snuffed out.

GALAXIES

Vast, rotating masses of stars, dust and gas, galaxies exist in huge numbers and are held together by gravity. Our solar system forms part of the galaxy known as the **Milky Way**, which consists of 100 to 200 billion other stars and measures around 100,000 light years across!

What is the Milky Way ❓

Astronomers estimate that our galaxy, the Milky Way is one of billions of other galaxies in the universe. The Milky Way contains about 200 billion stars, our Sun being one of them. It is 100,000 light years across. Our solar system is tiny in comparison to the size of the Milky Way, which can be seen clearly from Earth with a telescope.

Before telescopes, the Milky Way was viewed as a blurred, white streak across the sky. The ancient Greeks and the Romans named it 'a river of milk', and 'a road of milk', leading to its name today.

FACT FILE

The Sun is revolving around the centre of the Milky Way at a speed of 800,000 km/h, yet it will still take 200 million years for it to go around once.

The nearest galaxy to the Milky Way, Andromeda Galaxy, also a spiral, is about 2–3 million light years away.

Where is our solar system located inside our galaxy ❓

Located at the outer edge of the Milky Way, our solar system is 26,000 light years away from the centre of the galaxy. If you look towards the constellation of Sagittarius, you will be facing the central bulge of the galaxy. If you look towards the constellation of Cassiopeia, you will be facing the outer edge. If the Milky Way was reduced to 170 km in diameter, our solar system would be 2 mm in comparison!

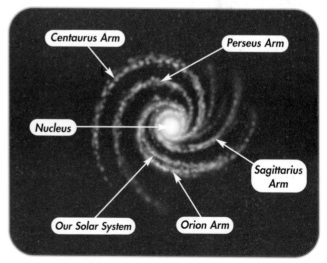

The Milky Way

What do galaxies look like

Galaxies can be split into different classes, depending on their shape. Our own Milky Way galaxy is a spiral shape. Spiral galaxies tend to contain high numbers of brighter, younger stars. Elliptical galaxies range in shape from spheres to ovals. They tend to consist mostly of older stars. Irregular Galaxies have little or no structure. They have probably become irregular in shape as a result of the gravitational force of other galaxies nearby.

Elliptical galaxies *Spiral galaxies* *An irregular galaxy*

An example of a spiral galaxy

What is Andromeda

The Andromeda galaxy, also known as Messier 31 (or M31), is only two million light years away and can be seen as a faint smudge of light in the night sky.

How many galaxies are there

There are billions of galaxies in the universe. No one knows for sure how many there are, but as technology advances, astronomers can see further away and detect objects they couldn't see before. In the early 1900s, Edwin Hubble discovered that our Milky Way was not the only galaxy in the universe.

The *Hubble Space Telescope* was launched in 1990, and orbits Earth, providing scientists with important information about galaxies in the universe.

This image from the Hubble Space Telescope shows galaxies outside our own

BLACK HOLES

Black holes sound like they could be something out of a science fiction story – objects so dense that nothing in the universe can escape from their gravitational pull. However, astronomers have been steadily building up evidence that black holes are not only real, but actually quite common in the universe.

What is a black hole ?

Black holes are regions of space where gravity is so strong that it gobbles up everything that comes near it – stars, planets, gases and even light! Black holes are invisible, but we know that they exist because we can see the stuff that is being sucked in to them!

The edge of a black hole is known as the 'event horizon' – the point of no return. As something is pulled inside, it is first torn apart by its immense gravitational force and then forms a flat rotating disc that spirals into the hole. Don't worry though, Earth is very far from any black holes!

How do we know they exist ?

Black holes can be detected only by their effects on the space surrounding them. As material gets closer and closer to a black hole it speeds up and bits start to smash together. This creates heat which we are able to detect. If the black hole is really large and has lots of debris in its disc, then it can reveal itself as one of the brightest objects in the universe – a quasar.

How does a black hole form ?

Black holes are thought to form from dead, or collapsed, stars. During most of a star's lifetime, nuclear reactions in the core generate an outward pressure that exactly balances the inward pull of gravity caused by the star's mass. When a star has burnt up its fuel, this outward pressure no longer exists

and the star falls in on itself, swallowed by its own gravity, with only the gravitational pull, or black hole, remaining.

The star collapses after all of its 'fuel' runs out

The star becomes very dense

The star is swallowed by its own gravity

What are wormholes

?

There is a theory that a black hole can form a tunnel through space called a wormhole. It is thought that if you entered a wormhole, you would travel through space and arrive in another universe. Although mathematically possible, wormholes would be extremely unstable and probably do not exist.

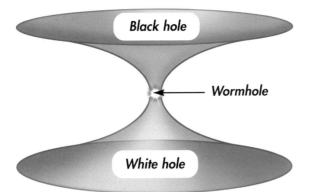

Black hole

Wormhole

White hole

FACT FILE

A feather would weigh several billion tonnes in a black hole!

If an astronaut 'fell' into a black hole, he or she would be stretched like a piece of spaghetti by the huge gravitational pull.

What is a supermassive black hole

?

Supermassive black holes are perhaps the most strange, destructive and terrifying objects in the universe! Billions of times bigger than our Sun, these powerful monsters may lurk at the centre of every galaxy – the Milky Way included. Scientists are beginning to believe that these forces of destruction actually help trigger the birth of galaxies and therefore are at the heart of the creation of stars, planets and all life.

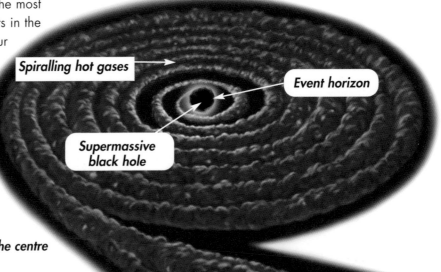

Spiralling hot gases

Event horizon

Supermassive black hole

A supermassive black hole could sit in the centre of our galaxy!

CELESTIAL BODIES

Along with the planets, stars and galaxies, there are comets, meteors and asteroids whizzing around the Sun. Members of NASA have devised something called the 'Torino Scale' to assess the damage celestial bodies could cause if they hit Earth. Fortunately, any serious collisions with the Earth are incredibly rare!

What is a comet ?

A comet is a chunk of icy and rocky material that releases gas and dust. If this icy chunk gets close to the Sun, it heats up and the material evaporates, creating the comet's coma, the surrounding cloudy atmosphere and its tail. Comets can be between 1 km and 50 km in diameter, whereas the comet's tail can be millions of kilometres long, even though the comet itself is small.

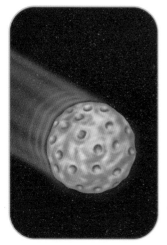

An icy comet ...

... is heated by the Sun

What is a meteor? Is it the same sort of thing ?

A meteor appears when a particle or chunk enters Earth's atmosphere from space. As the air friction heats the particle, it looks like a star shooting across the sky. Most meteors burn up before reaching Earth, however some are much bigger, and occasionally, a meteor which is too big to burn up reaches the Earth's surface. The meteors that reach the Earth are called meteorites, which vary in size. Most are very small, but the largest found was about 60 tonnes! At certain times of the year there are meteor showers. The meteors vapourise due to air friction, so the sky is filled with a shower of sparks!

A 'shooting star'

How many celestial bodies are there in our solar system

The LINEAR and NEAT telescopes in the US have discovered hundreds of thousands of tiny asteroids, mainly between Mars and Jupiter. Astronomers also know of over 1,000 comets, although most are too faint to be seen without a large telescope. The possibility of an asteroid hitting Earth is measured on NASA's Torino Scale. However, the chances of a collision which causes major damage is extremely low.

When can I see a comet

Comets can be seen from Earth when they pass close to the Sun, as the sunlight reflects off the gas and the dust. Every few years a comet becomes visible to the naked eye and, with binoculars, or a small telescope, you can see comets every year. Halley's comet becomes visible to the naked eye every 76 years as it nears the Sun.

Some comets are visible to the naked eye

What is the difference between an asteroid and a comet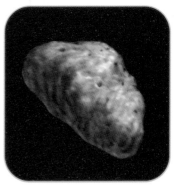

The main difference between asteroids and comets is what they're made of. While comets are made up of dust, ice and rocky material, asteroids are formed from rocky material and some metals like carbon and iron. This means that asteroids don't have a tail, as no ice evaporates from them.

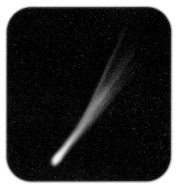

Asteroids reach 100 km across

Asteroids orbit the Sun, mainly between Mars and Jupiter. They range in size from 1–200 km across. The four largest asteroids known to astronomers are named Ceres, Pallas, Juno and Vesta.

The bright spot is called the nucleus

FACT FILE

Comets' tails don't point to the direction they're travelling, but are pushed away from the Sun by solar winds.

There is some evidence that William the Conqueror saw Halley's comet in 1066.

EARLY ROCKETS

The only vehicles powerful enough to carry people and equipment into space, rockets push themselves upwards and forwards using thrust. Without the rocket, we'd know very little about the things you are reading about in this book. We would be able to see certain planets and moons but we'd never be able to visit them.

Who invented the rocket ?

The first rockets may have been invented as early as the tenth century by the Chinese, but they were little more than flying 'fire-arrows' filled with flammable material. By the thirteenth century, however, primitive fireworks and rockets filled with gunpowder were in regular use. In the early 1800s, a military man Colonel William Congreve, based at Woolwich Arsenal, developed a range of rocket missiles which could be launched from special ships in battle.

Congreve rockets could be fired about 9,000 feet

Congreve rockets were built on long poles to make them easy to carry into battle

What was a liquid fuel rocket ?

The American Robert Goddard launched the first liquid fuel rocket on 16 March 1926. This technology would eventually get man to the moon and beyond. It used liquid oxygen and petrol as the propellant. Goddard's first rocket was over 3 m long, reached a height just over 12 m and a speed of nearly 100 km/h. By 1935 his rocket was nearly 5 m long and could climb to a height of over 2 km at a speed of nearly 1,000 km/h.

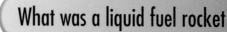

The first liquid fuel rocket

Dr Robert Goddard with an early liquid fuel rocket

Who was the first man to design a space rocket

Konstantin Tsiolkovsky (1857–1935) designed the first space rockets, although he never actually built any. He was a Russian mathematician and physicist who realised that liquid oxygen and liquid hydrogen would

provide the best rocket fuel. His theory, the 'Tsiolkovsky Formula', studied the relationships between rocket speed and gas pressure. It was not until the mid twentieth century that the first satellite was launched by the Russians.

Konstantin Tsiolkovsky

Who launched the first satellite into orbit

For over ten years, the Soviet Union and the United States of America were locked in a development battle known as the 'Space Race'. In 1957, the Soviet Union launched the first artificial satellite, *Sputnik 1*. It weighed 83 kg and orbited the Earth every 96 minutes for three months. The satellite helped to develop our understanding of the Earth's upper atmosphere. The successful launch shocked and amazed the world.

The Sputnik 1 *satellite*

Who was Wernher von Braun ?

Wernher von Braun was one of the most important pioneers of rocket development. He was employed under the Nazi regime to build V2 rockets and later transferred to NASA to help develop the Saturn rockets. He was the chief architect of the Saturn V launch vehicle, and is ultimately responsible for getting the first people to the Moon (see pages 52–54).

The V2 rocket was an impressive development and stood over 15 m tall. It was used to attack London during the war

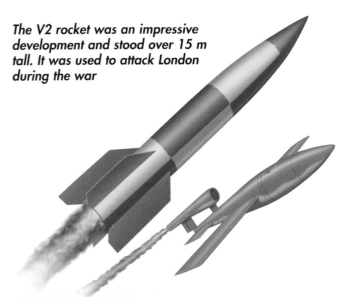

FACT FILE

NASA stands for National Aeronautics and Space Administration.

It was set up in 1958 as a worldwide body for space exploration and discovery.

MANNED ROCKETS

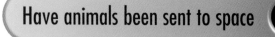

Once rocket technology had been developed, it was soon realised that if a missile could be sent across the world, rockets could be sent up into space. At the beginning of the 1960s, another race began between the Soviet Union and the United States of America to send the first man into space.

What was the first manned rocket

The first manned spacecraft was made in April 1961 when Yuri A. Gagarin orbited Earth in spacecraft, *Vostok 1*. The first manned spacecraft to orbit the Moon and come back was Apollo 8 in December 1968. *Apollo 11* landed on the Moon's surface on the 20 July 1969, commanded by Neil Armstrong.

Have animals been sent to space

Before humans were sent to space, American and Russian scientists sent animals in order to test the ability of launching live organisms into space. Monkeys, chimps, dogs and mice have all been carried to space, and have taught scientists a tremendous amount that could not have been learnt without them.

Who was the first man in space ?

On the 12 April 1961, Russian Yuri Gagarin became the first man in space. His spacecraft *Vostok 1* orbited Earth for 108 minutes. Less than a month later, NASA launched their first astronaut, Alan Shepard into space, but he did not orbit Earth. In February 1962, NASA launched John Glenn into space, where he orbited the Earth three times in five hours, reaching speeds of over 27,000 km/h.

Alan Shepard

Yuri Gagarin

Laika the dog was found on the streets in Russia before she became a space dog in 1957

How large are space rockets ?

The *Ariane 5*, which was launched on 21 October 1998, marked the beginning of a new style of heavyweight rocket. Its total height (including all stages) was 59 m, with a diameter of about 6 m. The *Saturn V* rocket that launched the Apollo spacecraft weighed over 3,000 tonnes. Nearly all that weight was fuel!

The Ariane 5 rocket

How are rockets sent to space ?

To send a rocket into space, it must be powerful enough to overcome Earth's gravity. Rockets work through a phenomenon called propulsion. Basically, the rocket is pushed forward because material is streaming out of the back of it. Most space rockets work through superheated gasses being pushed out of the back of them. These gasses are formed by burning fuel in the presence of oxygen, or another substance called an oxidiser that causes burning to happen. For a rocket to escape Earth's orbit, it must be travelling at a whopping speed of over 27,000 km/h!

The diagram shown right is a simplified version of the three-stage Saturn V moon rocket

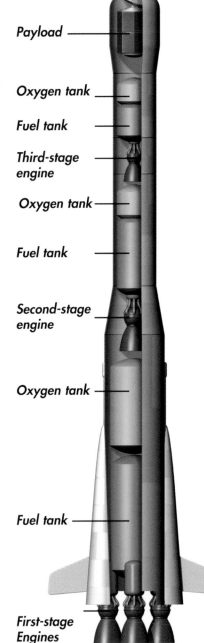

Payload

Oxygen tank

Fuel tank

Third-stage engine

Oxygen tank

Fuel tank

Second-stage engine

Oxygen tank

Fuel tank

First-stage Engines

Third stage:

The third stage is boosted into orbit with a smaller engine, which is discarded like stages one and two once the spacecraft is safely in orbit, or on the way to the Moon. The payload includes the passengers, cargo and equipment.

Second stage:

Liquid oxygen in the oxygen tank mixes with the fuel in the fuel tank of the second-stage engine, pushing the rocket to a higher altitude – around 185 km. Again, once used it is discarded.

First stage:

The first-stage rocket contains enough fuel to feed the engines, which provide enough lift for the huge weight of the rocket to escape gravity (i.e. – not fall back down again!). Once used the first one is discarded.

INSIDE A SPACECRAFT

Living in space may seem like fun. After all, there's no gravity to hold you down so your body floats. However, simple things we take for granted here on Earth, like taking a shower, or going to the toilet are a mission of their own inside a spacecraft. All astronauts must go through lots of preparation before they are considered ready for a journey into space.

How does space travel affect astronauts

The body's weightlessness in space means that bone and muscle can easily waste away, as there is little work for the muscles to do when the body weighs so little. Without gravity the spine starts to relax, and astronauts can easily be 5 cm taller at the end of a

mission. This height increase quickly goes away again after a few days back on Earth. To keep fit, astronauts eat a special diet and exercise regularly. More than two-thirds of astronauts suffer from motion sickness, although most recover after a few days in space.

Astronauts can become taller in space, as lack of gravity causes the spine to relax

Why are spacesuits essential

The atmosphere in space is very different to that on Earth, so a spacesuit provides astronauts with air to breathe when they go for a space walk outside the spacecraft. There is also no air pressure in space, so even if an astronaut could hold their breath, they could not go into space as their lungs would burst from the pressure inside their body! Spacesuits provide protection from the extreme temperatures in space too. Cold water was pumped around the suits used by the astronauts travelling to the Moon to keep them cool. The helmet on a spacesuit provides a protective dark visor to reduce the intense sunlight in space.

Inside the space shuttle

The Shuttle is divided into three sections: the flight deck, living quarters and lower deck. Everything, including the astronauts when they go to bed, must be strapped down!

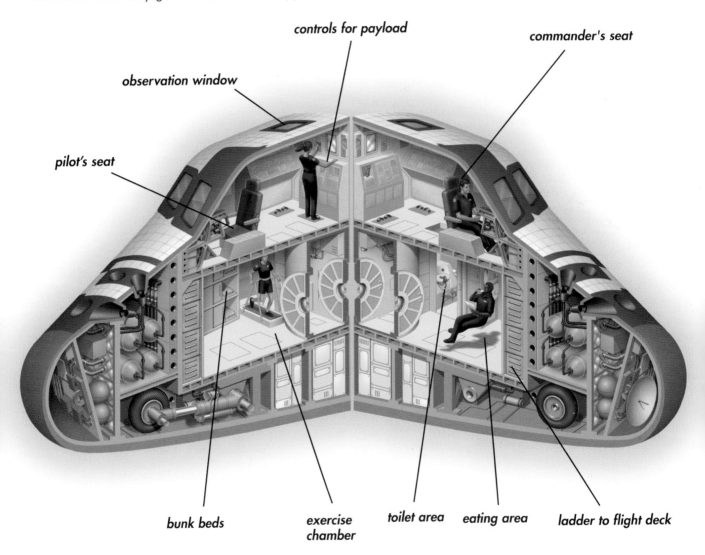

controls for payload

commander's seat

observation window

pilot's seat

bunk beds

exercise chamber

toilet area

eating area

ladder to flight deck

How do astronauts take a shower ?

Water is a very precious resource in space, so astronauts can go for days without a shower, just sponging themselves with damp cloths instead. On some spacecraft, a special 'shower' unit is fitted. The astronaut gets into the shower cylinder, shuts the door and then soaps up with a wet pad. The lack of gravity means that the water sticks to the body, so it has to be rubbed off. Some astronauts also use rinseless shampoo to save water.

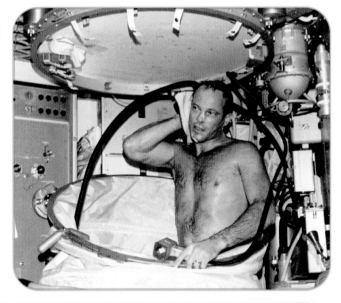

A space shower is very different than on Earth!

What do astronauts eat in space ?

In space, everything is weightless, which makes eating a normal meal nearly impossible. All the food, knives, forks and plates would just float around! To combat these problems, astronauts have special packs containing liquidised food which they can suck through a straw. Some of the meals like pasta are dehydrated so water is added to them through a straw, and then sucked up. Food like fruit can be eaten in its natural form.

FACT FILE

There is no definite explanation as to why food tastes differently in space. Some astronauts pack plenty of their favourite food, only to find they can't stand it after lift-off!

Astronauts often suffer from 'stuffy head' – they feel 'blocked up' around the upper half of the body because their blood flows in an upward direction!

Food must be freeze-dried to remove water to make it as light as possible. Bizarrely, the nutritional value remains nearly the same!

Grub's up!

Space food must be freeze-dried

MOON LANDING MISSIONS

The world watched with bated breath as NASA sent man to the Moon in 1969. From 1969 to 1972, six missions and twelve astronauts landed on the Moon, but there have been no more landings since. Several nations, including China and India, plan to send man back to the Moon within decades.

Who were the first people to land on the Moon

The first men to set foot on the Moon's surface were Neil Armstrong and Edwin 'Buzz' Aldrin, on 20 July 1969. It took them around four days to get there on the *Apollo 11* spacecraft, and they spent two and a half hours on the surface, collecting soil samples and taking photographs. While on the Moon, they spoke to US President Richard Nixon in the White House and planted a US flag. When Armstrong stepped onto the Moon, he uttered the phrase, 'One small step for man, one giant leap for mankind'.

What were the Apollo landings

The Apollo missions were designed to send humans to the Moon, where they could collect information and conduct scientific research, and return them safely to Earth. Between 1969 and 1972, Apollo spacecraft landed on the Moon six times, visiting different areas each time. Astronomers gave these landing areas different names, like the 'Sea of Tranquillity' and 'Hadley Rille'. The astronauts on *Apollo 15* travelled 25 km in a Land Roving Vehicle (LRV), an electrically powered four-wheel drive vehicle. An attempt by *Apollo 13* to land on the Moon failed due to an onboard explosion, but the spacecraft did manage to return safely to Earth.

An estimated audience of 700 million people watched the event on television.

Will man return to the Moon in the future

NASA is planning to send a manned mission to the Moon by 2013. India and Russia have also expressed interest in sending manned missions to the Moon, and China are aiming to send people to the Moon within the next decade. The United States and Russia plan to establish manned bases on the Moon. They even hope that this can be achieved as soon as 2024–2032.

If a permanent base can be established on the Moon, it could be used as a testing ground for space equipment and technology, and lead to the discovery of new energy supplies.

Why were the Moon landings so important

A lot was learnt from the manned missions to the Moon – enough information was gathered for scientists to work on for decades! Tonnes of lunar rock was brought back to Earth, and hundreds of photographs were taken. During the missions, astronauts left experiments on the surface of the Moon, which sent data back to

Earth for many years. The Earth's magnetism was measured, along with the distance between Earth and the Moon. The missions proved that human beings could live and work in space without suffering ill effects.

The distance between the Earth and the Moon was measured

Tonnes of lunar rock was brought back to Earth

Could we go on holiday to the Moon one day

Scientists believe that eventually people will be able to travel into space much like they travel on aeroplanes today. Individuals have travelled into space before. In 2001, Dennis Tito, an American businessman, paid around £10 million to travel aboard the Russian spacecraft *TM-32*. In 1986, Christa McAuliffe, an American teacher, was killed on board the space shuttle *Challenger*, when it broke apart, leading to the deaths of all its crew. These days, it is possible to pay for a flight to space with the Russian Space Agency, costing around £15 million!

Were the Moon landings faked

Some people, called conspiracy theorists, argue that the Moon landings didn't happen at all! They claim that the video footage and photographs taken during the missions were faked by NASA, who continue to deceive the public to this day. Some of these claims can be discredited by three reflectors left on the Moon by *Apollo 11, 14* and *15*. Today, anyone on Earth with an appropriate laser and telescope system can bounce laser beams off these devices, proving that they were indeed left on the Moon during these missions as NASA claimed they were.

RETURN TO EARTH

Once astronauts have gathered all the evidence that they need, taking samples or images from space, they need to return to Earth. (This could be after a considerable period – Russian Valeri Polyakov, spent 438 days in the *Mir Space Station* in 1994–95.) Returning to Earth needs to be planned very carefully to ensure that the craft descends safely when it pushes through Earth's atmosphere.

How does re-entry happen ❓

Returning to Earth from space requires a great deal of skill and planning. The trick is to enter Earth's atmosphere at just the right angle to slow the spacecraft down safely, without using up huge amounts of fuel. The pilot receives instructions from 'space station control' on Earth to help them navigate Earth's atmosphere. Earth's atmosphere also acts as a drag, slowing down the shuttle.

'Space station' control on Earth

A deck in a spacecraft

Why is re-entry difficult ❓

If the craft enters at too shallow an angle it will bounce off the atmosphere. If it enters at too steep an angle it will burn up and hit the ground at great speed. By entering the atmosphere at the correct angle, the spacecraft will be able to reduce the heat and resistance generated, eventually landing safely.

Retro-rockets fire to slow the Shuttle down and the spacecraft turns so the heat shield faces the atmosphere

The Shuttle usually starts to re-enter the atmosphere on the dark side of the planet

As the Shuttle approaches, it becomes a huge glider, landing with no engines, at over 320 km/h

Below 1,600 km/h, the Shuttle starts to land like a normal aircraft, weaving in S-shaped curves to further reduce speed

The Shuttle enters the densest parts of the atmosphere at about 12,000 m above Earth

As the Shuttle has no reverse thrust jet engines, a parachute and wheel brakes finally bring it to a stop

The process of returning to Earth takes about one hour

What happened to the space shuttle *Columbia*

On 1 February 2003, the space shuttle *Columbia* burned up on re-entry, killing all seven astronauts. A large chunk of insulation foam, which fell off the fuel tank at lift-off, damaged the shuttle's heat-resistant tiles, causing the spacecraft to overheat and break up on re-entry. This wasn't the first tragic shuttle accident. In 1986, seven astronauts were killed when the shuttle *Challenger* developed a fault with its boosters.

What are heat-resistant tiles

A space shuttle is covered with around 20,000 tiles that can withstand up to four times more heat than they encounter upon re-entry. The tiles are made of a special material that can be red hot on one side but cool on the other. NASA space shuttles are designed to be reusable. When they re-enter Earth's atmosphere many of the tiles are damaged and need to be replaced. It is far easier to replace small tiles than replace one huge heat shield.

The Space Shuttle Columbia *blasts off*

The ceramic heat tiles are reusable but take a real bashing when they re-enter the atmosphere

FACT FILE

Some shuttles land on a runway, in a similar way to a plane. They have a lifting body design and swept-back wings. When a shuttle 'touches down' this way, the descent rate is about seven times steeper than that of an aeroplane!

SATELLITES

A satellite is any object that orbits or revolves around another. In addition to the Moon (a natural satellite), thousands of man-made satellites also orbit Earth. They are used for many different purposes, including satellite television, phone calls, radio transmissions, Internet connections, weather forecasting, scientific research and surveillance.

How high are satellites ?

Many scientific satellites are only 300–400 km above Earth and orbit in under two hours. As these satellites are relatively close to Earth, they require little fuel and are cheap to launch. Major communications satellites are put at a huge distance of 36,000 km from Earth. At this distance, an orbit takes exactly one day to complete. Satellites have also orbited the Sun, Moon, asteroids and some planets.

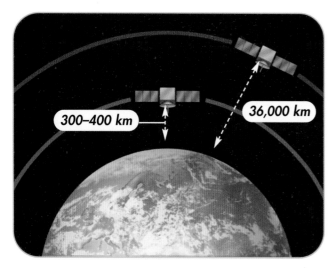

300–400 km   36,000 km

How many man-made satellites are out there ?

Man-made satellites are also called 'artificial satellites'. Since the late 1950s, thousands of satellites have been launched. Many of the early satellites have fallen back to Earth and burned up in the atmosphere. There are also thousands of items of 'space-junk', such as rocket boosters and fuel tanks that have not burned up in orbit. Today, there are about 3,000 useful satellites and 6,000 pieces of 'space junk' orbiting Earth.

Why do satellites burn up ?

Many satellites do not remain in their orbits, but gradually return to Earth. As a satellite loses altitude (height) it enters denser regions of the atmosphere, where friction between the satellite and atmosphere generates a great deal of heat. The air around the satellite becomes so hot that pieces of the satellite break into smaller pieces, eventually burning up or disintegrating. Some satellite components can survive the re-entry heating, crashing down to Earth at tremendous speeds!

What do satellites actually do

The satellites that orbit Earth perform many different jobs. There are telephone, TV and radio communications satellites, military spy satellites, weather satellites and satellites studying the Sun and distant objects in the universe. There is also the *Hubble Space Telescope* which is one of the largest objects orbiting Earth. The biggest object is the *International Space Station.*

We use satellites to communicate

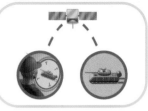

Some are used by the military

high gain antenna – provides the satellite with radio control

'space junk' – debris from broken up satellites

this contains propellant and all the instruments required to send information to Earth

FACT FILE

The *International Space Station* is a manned space station which orbits Earth.

It is a joint project by the US, Russia, Japan, Canada and several European countries.

solar panels

satellites may also contain x-ray detectors

FUTURE MISSIONS

So what does the future of space travel hold? In the last century, humans have sent manned missions to the Moon and satellites to distant worlds – things that early astronomers could only dream about. However, if we are ever to send manned missions to the distant planets of our solar system and beyond, some mighty challenges need to be overcome and major new technologies developed.

Should we use humans or robots ?

Although sending human beings into space was a great achievement, small robotic explorers appear to be the most efficient way of exploring space. The advances in computing power in recent years means that lightweight robots can land on planets like Mars and explore the surface. The biggest advantage to using machines is that they can be replaced if they are destroyed, whereas human beings cannot be replaced.

1 year

1 year

The Mars Rover *Explorer*

Will humans ever visit Mars ?

For decades, governments and space agencies have considered sending manned missions to Mars. In 2007, NASA hinted that they may be able to launch a human mission to Mars by 2037. The European Space Agency has the long-term vision of sending a human mission to Mars by 2030. Some scientists have criticised these plans, arguing that manned missions to Mars would be too expensive and that funding would be better spent developing other space technology such as robots.

How can we travel to distant planets more quickly

One of the most likely methods of powering spacecrafts to distant planets would be to use nuclear power. Nuclear rockets are much more powerful than our existing chemically-fired launchers, and rockets run by nuclear fission are fuel efficient and very light. They could reach Saturn in three years instead of seven! Missions would become easier, as the need to carry food, fuel and oxygen would be reduced.

In 2003, NASA established Project Prometheus, a project that develops nuclear-powered systems for long distance missions. In 2005, the project faced an uncertain future, and was likely to be reduced to a low cost research project. However, the use of this type of fuel is highly controversial. The rockets could release nuclear waste into the atmosphere.

Where will we go next ?

FACT FILE

Over 3 million people had their names burned onto an electronic disk, attached to the Exploration *Rover* lander that trawled the surface of Mars.

During the past forty years, there have been huge changes and developments in space flight. Most astronomers would like to think that at some point mankind will spread out into the solar system and explore other galaxies. However, as there are no really fast spaceships, it would take years to get to the closest planets in our solar system. Rockets are propelled by controlled explosions and not much has changed in the last 40 years. To travel further into space, new ways of powering our spacecraft need to be developed.

The next generation of space exploration needs faster spaceships.

# GLOSSARY

Astronaut

A person trained for travelling in space.

Astronomer

Those who study stars, planets and the universe.

Astrophysics

The branch of astronomy that is concerned with the physical side of stars and planets.

Atmosphere

The layer of gas surrounding Earth or other planets.

Big bang

A cataclysmic explosion that scientists believe created time and space.

Big crunch

What might happen if the universe stops expanding and collapses on itself.

Billion

A thousand million (1,000,000,000).

Black hole

A region of space caused by the collapse of a star, so dense that neither matter or light can escape.

Cassiopeia

An easily-spotted W-shaped constellation near the Pole star.

Celestial

relating to the sky or outer space

Comet

Pieces of ice and dust which orbit the Sun.

Constellation

Any of the 88 groups of stars as seen from Earth, named by the Greeks after mythological people, objects or animals.

Corona

The very hot outer layer of the Sun's atmosphere.

Ellipse

A shape like a flattened circle.

Equator

The name for the imaginary band around the middle of Earth that splits it into two hemispheres, the north and the south.

Evening Star

Another common name for Venus.

Fission

The splitting of the centres of heavy atoms into lighter ones.

Fusion

The combining of lighter elements into heavier ones.

Galaxy

A collection of billions of stars held together by gravitational attraction.

Great Red Spot

A long-lived feature on Jupiter's surface, south of its equator, which has survived for hundreds of years.

Hemispheres

The two halves of the globe, as divided by an imaginary line around the middle called the equator.

Hydrosphere

The water on, or around, the surface of a planet.

International Space Station.

A permanently manned satellite constructed between 1998 and 2001 for space research.

Kelvin

A temperature scale used by astronomers, in which the lowest possible temperature is called 'absolute zero'.

Lunar-Roving Vehicle

An electronically powered four-wheel drive vehicle that can explore planet's surfaces.

Light year

The distance that light travels in one year. It is equal to just under 10 trillion kilometres!

LINEAR

Also NEAT – Telescopes which discover asteroids and comets.

Lunar eclipse

When Earth passes between the Sun and the Moon.

GLOSSARY

Magnetosphere

A magnetic field around the Sun and certain planets.

Magnitude scale

The scale on which objects are measured for their brightness.

Mariner

A series of American space probes which visited Mercury, Venus and Mars.

Messier

A catalogue named after an eighteenth century astronomer which identifies galaxies and nebulae etc. by their number i.e. M31 (Andromeda Galaxy).

Meteor

A very small iron or rocky particle that has entered Earth's atmosphere. Also called a shooting star.

Milky Way

A spiral galaxy containing billions of stars, our solar system and Earth.

Miranda

One of Uranus' largest moons, with a unique surface.

NASA

The National Aeronautics and Space Administration is an agency of the United States government responsible for the nation's public space program.

Nova

When a star undergoes an eruption and suddenly becomes much brighter for a short period.

Observatory

A building specially designed to look at astronomical objects.

Olympus Mons

A massive volcano on the surface of Mars.

Orbit

The regularly repeated path of a Moon, spacecraft etc. around a star or planet.

Photosphere

The area of the Sun that we look at, face-on.

Radiative zone

The heat from the Sun's core is passed through this zone, and the energy transfer begins. It then reaches the less dense area of the convective zone where it rises and starts to reach the atmosphere.

Rings

Any of the thin, circular bands made from small components that orbit something larger i.e. Saturn.

Satellite

A satellite is an object that rotates around or orbits another object. They can be man-made like the communications satellites that orbit Earth; or natural like the Moon that orbits Earth.

Sea of Tranquillity

A smooth area of ancient lava on the surface of the Moon – made famous by the *Apollo 11* landing.

Solar eclipse

When the Moon comes between the Sun and Earth.

Solar flare

A sudden brightening near a sunspot.

Solar System

The system containing the Sun and other bodies in its gravitational field, including the Moon and nine known planets.

Sunspot

A dark blemish on the solar surface that is caused by the Sun's magnetic field.

Sunspot cycle

The eleven-year cycle that sees the rise and fall in the number of sunspots.

Supernova

When a star runs out of fuel, it becomes unstable and appears 100 million times brighter for a few days before 'dying'.

White dwarf

A small, very dense star that has come to the end of its life.

INDEX

Illustrations by Stephen Sweet.
Photographs reproduced by kind permission of NASA.

FOCUS  N

Collect them all!

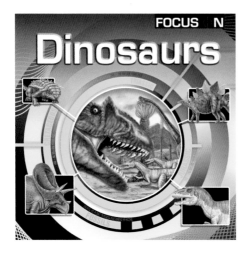

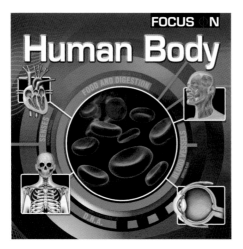